UNDERWATER WEAPONS

ANIMAL WEAPONS

Lynn M. Stone

The Rourke Press, Inc.
Vero Beach, Florida 32964

© 1996 The Rourke Press, Inc.

All rights reserved. No part of this book may be reproduced or utilized in any form or by any means, electronic or mechanical including photocopying, recording or by any information storage and retrieval system without permission in writing from the publisher.

PHOTO CREDITS
© Marty Snyderman: cover, title page, p. 4, 7, 10; © Tom Campbell: p. 13; © Lynn M. Stone: p. 8, 12, 15, 17, 21; © Frank S. Balthis: p. 18

Library of Congress Cataloging-in-Publication Data

Stone, Lynn M.
    Underwater weapons / Lynn M. Stone.
      p. cm. — (Animal Weapons)
    Includes index
    Summary: Describes the tentacles, teeth, claws, and other parts of their bodies that undersea creatures use to hunt and to protect themselves.
    ISBN 1-57103-162-6
    1. Animal weapons—Juvenile literature. 2. Marine animals—Juvenile literature. [1. Marine animals. 2. Animal weapons.] I. Title II. Series. Stone, Lynn M. Animal weapons.
QL940.S748 1996
591.53—dc20   96–8995
                                                     CIP
                                                       AC

Printed in the USA

TABLE OF CONTENTS

Underwater Weapons	5
Tentacles	6
Jellyfish Tentacles	9
Fish Teeth	11
Mammal Teeth	14
Claws	16
Spines	19
Muscles and Snouts	20
Stun Guns?	22
Glossary	23
Index	24

UNDERWATER WEAPONS

The underwater world is a dangerous place for the animals that live there. Large animals prowl the sea looking for smaller animals to kill and eat.

Hunters and hunted alike have developed ways to survive—at least some of the time.

Weapons are an important means of survival. Both **predators** (PRED uh turz), the hunters, and **prey** (PRAY), the hunted, depend upon them.

Lethal (LEE thul), or deadly, weapons help animals kill other animals. They also help animals defend themselves.

Don't be fooled by its beauty! The liontish is one of several fish with poisonous spines

TENTACLES

Several underwater animals have what look like snakelike arms. They're not really arms. They're boneless **tentacles** (TEN tuh kulz). They help the octopus and their cousin the squid catch prey.

The giant squid has the most remarkable tentacles. They can be 60 feet long. The animal weighs up to 1,000 pounds.

Octopus and squid tentacles reach out, grab, and pull such prey as crabs, worms, and fish to the animals' mouths.

Tentacles of squid grab hold of prey

JELLYFISH TENTACLES

The tentacles of sea anemones, corals, and jellyfish are delicate. The tentacles of some jellyfish look like long threads. The tentacles of anemones and coral are shorter. They are more like fingers and flowers.

The anemones, corals, and jellyfish are soft, simple animals without eyes. Their tentacles find prey by touch. Tentacles of these animals release tiny, stinging darts that can be lethal to prey. Some jellyfish stings are even lethal to people.

Fragile, fingerlike tentacles of sea-anemones sting prey

FISH TEETH

Teeth are the weapons of many animals above and below the water surface.

Some of the sharpest teeth in the animal kingdom belong to fish. Many **species** (SPEE sheez), or kinds, of sharks have several rows of sharp teeth. A great white shark has teeth more than two inches long!

Barracudas and some eels also have extremely sharp teeth. These fish prey largely upon other fish.

A great white shark's big, powerful body and sharp teeth make it one of the ocean's most fearsome predators

Sea stars use their "arms" to pull open shells of mussel clams

The scorpion fish's lacy look disguises its poisonous spines

MAMMAL TEETH

The largest whales in the sea are toothless. The 60-foot sperm whale, however, has a full set of cone-shaped teeth. Sperm whales eat fish and giant squid.

The killer whale also has large, cone-shaped teeth. Most killer whales are fish eaters, but a few kill seals.

Seals are toothy, too. Most of them catch fish. Leopard seals often feed on penguins that they grab with their teeth.

Most dolphins and other toothed whales use their cone-shaped teeth to snag fish and squid

CLAWS

Lobsters and crabs are two of the sea world's best-known crusty creatures. Both have hard-shelled claws.

Lobster and crab claws aren't like the claws of bears or big cats. Lobster and crab claws have upper and lower sections. The claws snap open and shut like pliers.

Lobsters and crabs use their claws both to defend themselves and to kill other sea creatures.

The shelled claws of a northern lobster help it survive in the sea world

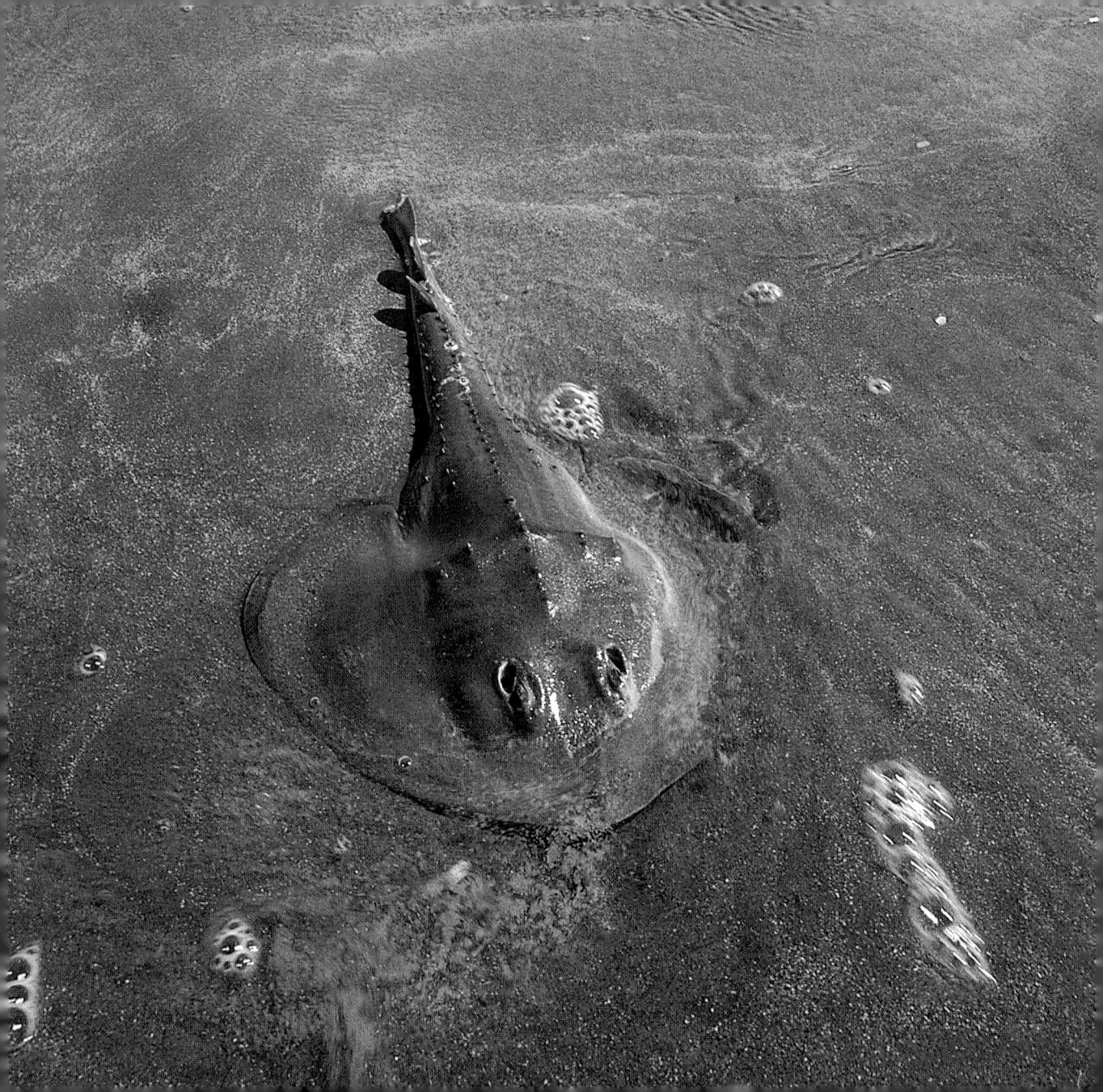

SPINES

Many marine animals have sharp needles called spines. Spines help animals protect themselves.

Several kinds of sea urchins are covered by spines. Some animals, like horned sharks and stingrays, have only one or two spines. The sting ray's barbed spines are on its tail.

Sting ray spines and the spines of many other fish are very poisonous. Swimmers sometimes step on the spines of sting rays resting in the sand under shallow ocean water.

Spines of the sting ray put poison into the wounds they cause

MUSCLES AND SNOUTS

A snail doesn't appear to have any weapons. It looks quite helpless.

The tulip snail and its cousins, however, are predators. They use their "foot" muscles to open clams. The snail's long snout, called a **proboscis** (pro BAHS iss), attacks the prey's soft parts.

The cone shell uses its proboscis to shoot a sharp, toothlike object at fish. The tooth not only injures the fish, but it carries the cone shell's poison as well.

The tulip shell travels on a "foot" that also serves as a weapon against the shelled animals it pries open and eats

STUN GUNS?

Do certain whales shout fish to death?

Scientists know that certain toothed whales locate fish in deep, dark water by sound. When a whale's undersea calls strike an object, they make echoes. By "reading" the echoes, a whale knows where its prey is.

Some scientists believe these whales make loud, lethal sound waves. The sounds might be powerful enough to injure or even kill fish, these scientists say.